Software's Sentinel

Safeguarding Code's Integrity Through Rigorous Testing And Validation

Lacy Zamorano

Contents

1 Introduction to Software Testing

Software testing isn't just about finding bugs; it's about building confidence. It's the crucial process that ensures your software functions as intended, performs reliably, and meets user expectations. Think of it as a safety net, catching potential problems before they reach your users, saving time, money, and reputation. Neglecting testing is like building a house without checking the foundation – a recipe for disaster. The cost of fixing a bug discovered late in development is exponentially higher than addressing it early on. This is why thorough testing is not just good practice, but an absolute necessity.

Testing methodologies provide a structured approach to finding and fixing software defects. Different methodologies suit various project needs and complexities. Waterfall, for instance, employs testing at distinct stages of development, while Agile integrates testing throughout the iterative development process. Choosing the right methodology depends on factors like project size, team structure, and the desired level of rigor. Understanding these differences is key to selecting the most effective strategy for your project. Remember that a well-defined testing plan is crucial for success, no matter what methodology you choose. It is essential to outline clear objectives and milestones from the start and to frequently assess progress toward meeting them.

A comprehensive testing strategy includes various levels. Unit testing focuses on individual components, ensuring they function correctly in isolation. Integration testing verifies the interaction between different components, while system testing evaluates the entire system as a cohesive unit. Each level plays a vital role in uncovering different types of defects. A well-rounded approach encompasses all of these levels, addressing potential issues at every stage of development, and should be approached methodically and planned from the early stages of your project. It's easy to overlook the necessity of testing at every stage, but overlooking it can lead to much greater issues later in development.

Beyond functional correctness, you need to consider non-functional aspects like performance, security, and usability. Performance testing measures response times and resource usage under different loads. Security testing identifies vulnerabilities that could be exploited by malicious actors. Usability testing ensures the software is intuitive and user-friendly. Neglecting any of these areas can lead to a subpar user experience, security breaches, or system instability. It's important to carefully evaluate each requirement and to determine a good strategy on how to test them, making sure to not overlook any requirement.

Test automation plays a crucial role in modern software development. Automating repetitive testing tasks improves efficiency, reduces human error, and enables more frequent testing. Tools and frameworks exist to automate various testing levels, from unit tests to UI tests. Investing in automation can significantly improve the quality and speed of your software development process. Learning to implement these testing frameworks and tools is important, but also deciding which tools you want to use requires thinking carefully about the type of application you are testing, and what is available to you.

Finally, effective software testing involves more than just technical skills. It demands meticulous planning, attention to detail, and a systematic approach. This systematic approach also includes properly documenting all testing procedures and results, which not only ensures that the process is well understood, but also can be used for better tracking purposes and can be used as references for future testing. Strong communication and collaboration between testers, developers, and stakeholders are essential for identifying and resolving issues effectively. Building a strong team around this effort is also important.

The process of software testing is ongoing, it's not a one-time activity; it's an integral part of the entire software development lifecycle. Continuous testing and integration help identify and resolve issues early, preventing them from escalating into larger problems. It ensures consistency and allows for frequent evaluations to prevent larger issues from happening in the future. Consistent testing also helps avoid problems happening in the future. This ongoing testing and continuous integration approach is essential to build high-quality and reliable software.

1.1 The Importance of Testing

Software testing isn't merely a final step; it's the lifeblood of reliable software. Think of it as a rigorous quality control process woven into every stage of development, preventing costly mistakes and ensuring user satisfaction. Ignoring testing is like building a house without a foundation—it might stand for a while, but it's destined to crumble under pressure.

Effective testing is proactive, not reactive. It's about identifying and fixing bugs early, before they become major headaches. This saves time and resources, ultimately resulting in faster development cycles and happier clients. Imagine finding a critical flaw just before a major release—the cost of fixing it then is exponentially higher than addressing it during early development stages.

Early testing minimizes such expensive surprises.

Consider the different types of testing. Unit testing focuses on individual components, ensuring each piece works as expected. Integration testing then checks how these components work together, revealing flaws in their interaction. System testing is the final, comprehensive assessment, verifying the whole system meets requirements. Each level builds upon the previous one, providing a layered defense against defects.

Testing isn't just about finding bugs; it's also about validating the software's functionality. Does it meet the user's needs? Is it user-friendly and intuitive? Performance testing assesses speed, responsiveness, and stability under various loads. Security testing is equally crucial, identifying vulnerabilities that could expose sensitive data or compromise system integrity. Thorough validation helps ensure the software is not only functional but also robust, secure, and user-friendly.

Test automation plays a critical role in modern software development. Automating repetitive tasks frees up developers to focus on more complex aspects of testing. Automated tests can be run frequently, providing rapid feedback and catching errors early. This is particularly crucial in Agile development environments, where rapid iterations are the norm. Imagine the efficiency gains of having automated tests run nightly, providing immediate notification of any issues that arise from the day's code changes.

Documenting your testing process is paramount. Detailed test plans, test cases, and reports provide a clear record of what was tested, how it was tested, and the results. This documentation is essential for troubleshooting issues, improving future testing efforts, and ensuring compliance with quality standards. Imagine having to debug a system without clear documentation of its testing history—finding the source of a bug would become a near-impossible task. Consistent documentation provides valuable insight and simplifies debugging.

While automation is valuable, human expertise remains crucial. Experienced testers bring a critical eye to the testing process, identifying subtle issues that might be missed by automated tools. Their judgment and experience are invaluable in ensuring comprehensive testing and interpreting test results. Automated tools are fantastic aids, but human intuition remains an indispensable part of thorough testing.

Beyond technical skills, a successful tester needs strong communication and collaboration skills. Testers frequently interact with developers, project managers, and stakeholders, effectively communicating test findings and collaborating to resolve issues. Strong communication skills are essential for conveying technical information clearly and concisely, preventing misunderstandings and facilitating productive discussions. Testers serve as vital communicators, bridging the gap between the technical and nontechnical aspects of the development process. Their feedback can be crucial for improvements before a software is released.

Finally, ethical considerations are crucial in software testing. Testers have a responsibility to ensure the software they test is safe, reliable, and meets ethical standards. This involves identifying potential biases in the software, ensuring it doesn't perpetuate harmful stereotypes, and contributing to the development of responsible and ethical software practices. A tester's duty is not only to find bugs but also to ensure the software aligns with ethical principles, ensuring a positive impact on society. The ethical implications of software and its consequences must be considered during testing phases.

1.2 Testing Methodologies

Choosing the right testing methodology is crucial; it significantly impacts project success. A poorly chosen approach can lead to missed bugs, increased costs, and ultimately, a flawed product. Understanding your project's specific needs and constraints—budget, timeline, complexity, and the risk tolerance—is paramount before selecting a methodology. Careful consideration of these factors will prevent you from wasting precious resources on an inappropriate strategy.

Agile methodologies, for instance, favor iterative testing, emphasizing frequent feedback loops and adaptation throughout the development lifecycle. This approach perfectly suits projects with evolving requirements or those needing quick iterations. Conversely, Waterfall, with its sequential nature, is more suitable for projects with well-defined, unchanging requirements, where thorough upfront planning is key. Remember, selecting the wrong methodology could be like trying to build a skyscraper with only toy bricks.

Consider the V-model, a structured approach that mirrors the development lifecycle. Each phase of development has a corresponding testing phase. This meticulous mapping enhances traceability and simplifies the identification of defects. While highly structured, it lacks the flexibility of Agile, making it unsuitable for rapidly changing project needs. Choosing this method demands a clear and stable project scope; any major change during development will disrupt the carefully balanced stages and can prove exceptionally problematic.

Spiral testing balances iterative development with risk management. Each iteration involves prototyping, testing, and risk assessment, allowing for adjustments based on emerging concerns. This risk-focused model is particularly useful for complex projects where unexpected issues can drastically impact outcomes. It needs experienced team members, as a lack of expertise could lead to poor risk assessments and, consequently, inefficient solutions.

Then there's the incremental model, building the software in small, manageable increments. Each increment undergoes thorough testing before being integrated with the next. This modular approach simplifies testing, allowing for early detection of defects and making maintenance easier. It's ideal for large, complex projects where breaking down the task into smaller, manageable components facilitates development and testing. However, this approach requires careful planning and coordination across the teams involved.

The Big Bang approach, on the other hand, involves testing the entire system only once it's completed. While seemingly simple, it's inherently risky. Debugging becomes significantly more challenging with this approach, potentially leading to significant delays and cost overruns. Only use this method for the smallest, simplest applications where the risk is deemed acceptable. Even then, it's usually far less efficient than other testing methodologies.

Ultimately, the best testing methodology depends on several factors, including the software's complexity, budget, and time constraints. Experienced professionals should carefully weigh these factors against the benefits and drawbacks of each approach. It's not just about choosing a methodology; it's about choosing the methodology that best protects your software and fits your project's unique challenges. Selecting the appropriate method is an investment in the quality and success of your software. A thorough understanding of each methodology enables informed decision-making, ensuring that the chosen path aligns with project goals and resource limitations. Don't rush this critical decision; careful analysis will save you valuable time and resources in the long run. Prioritize a well-informed decision over a hasty one; the quality of your software depends on it. The goal isn't simply to test the software; it's to rigorously validate its functionality and security, protecting it from vulnerabilities and ensuring user trust. This requires expertise and planning, and a chosen methodology will only support such goals.

2 Unit Testing Essentials

Unit tests are the cornerstone of robust software. They verify the functionality of individual components, ensuring each piece works correctly before integrating them into larger systems. Think of it as meticulously checking the gears of a clock before assembling the entire mechanism; a single faulty gear can disrupt the entire timekeeping process. Similarly, a flawed unit can cripple an entire application.

Writing effective unit tests requires careful planning and execution. Begin by identifying the smallest testable units within your code, often individual functions or methods. These units should have clear, well-defined inputs and expected outputs. The tests themselves should isolate these units, minimizing external dependencies as much as possible. This isolation ensures that a failed test truly points to a problem within the unit being tested, not some unforeseen interaction with another part of the system. Using mocking frameworks, you can simulate interactions with external systems, databases, or other components, allowing for controlled testing without relying on unpredictable external factors.

Test-Driven Development (TDD), a powerful technique, flips the traditional development approach. Instead of writing code first and then testing it, in TDD you begin by writing a failing unit test that defines the desired functionality. Only then do you write the code to pass that test. This iterative process ensures that you only write code that meets specific requirements, promoting cleaner, more maintainable code. Moreover, the tests themselves serve as comprehensive documentation of the intended behavior of each unit. This process naturally leads to more resilient code, as each component is thoroughly validated before being incorporated into larger systems. The benefits extend beyond the initial development phase, making future modifications and maintenance less error-prone.

One common pitfall is writing tests that are too complex or coupled to external factors. Overly complicated tests are difficult to debug and maintain. They may be fragile, breaking easily with minor code changes unrelated to the actual functionality being tested. Poorly designed tests can mask underlying problems, leading to a false sense of security. Avoid testing implementation details; instead, concentrate on verifying the expected behavior of the unit. A comprehensive suite of well-designed unit tests will significantly improve the software's quality and maintainability.

Remember that tests should be independent, each verifying a specific aspect of the unit's functionality, ensuring that a failure in one test doesn't cascade into others. Strive for simplicity; a concise and well-focused test is always preferable to a lengthy, tangled one.

Another critical aspect is the use of appropriate testing frameworks. These frameworks provide structures and tools to streamline the testing process, making test creation and execution more efficient and manageable. Different programming languages and development environments offer diverse testing frameworks, each with its strengths and weaknesses. Carefully selecting a framework appropriate for your project and understanding its features is crucial for achieving optimal testing efficiency. Consider your project's scale and complexity when choosing a framework, opting for a framework that adequately supports the project's needs without adding unnecessary complexity. These frameworks offer features that simplify testing, including test runners, assertion libraries, and mocking capabilities.

Comprehensive unit testing significantly reduces the likelihood of bugs reaching later stages of the development process. Early detection of defects is crucial for cost-effective software development. The cost of fixing a bug detected during unit testing is significantly lower compared to identifying it during integration or system testing. This cost-effectiveness makes unit testing an invaluable part of any robust software development methodology. It allows developers to quickly pinpoint errors, accelerating the debugging process and reducing the overall time required for software development. This time saved can translate directly into faster time-to-market and better resource allocation for projects. Remember that the investment in unit testing always pays off in improved software quality and reduced development costs.

Finally, remember that unit testing is an iterative process. As your software evolves, so should your tests. Regularly reviewing and updating your tests to reflect changes in the code ensures they remain accurate and relevant. Outdated tests can provide a false sense of security, masking potential issues. Maintaining up-to-date tests is crucial for continuously safeguarding the integrity and reliability of your software. This proactive approach ensures that your tests always provide accurate feedback on the state of your code, and it remains an essential component of any ongoing maintenance and improvement strategy for your projects. By diligently following these guidelines, you'll build a reliable and maintainable software system, thereby reducing the likelihood of encountering unforeseen problems and ultimately strengthening the quality of your work.

2.1 Writing Effective Unit Tests

Effective unit tests are concise, focused, and easily understood. They should test one specific aspect of your code's functionality, isolating it from external dependencies to pinpoint the source of any failures. Imagine testing a function that calculates the area of a circle; a good unit test would only focus on the accuracy of the area calculation, not on the validity of the radius input. This isolation simplifies debugging and maintenance considerably. Think of it like diagnosing a car engine problem: you wouldn't start by replacing the entire engine; you'd systematically check individual components.

Properly structured unit tests enhance code readability. Well-named test functions clearly convey their purpose. For example, test_calculate_circle_area_positive_radius is far more descriptive than test_1. Consider using descriptive variable names within your tests as well. This clarity significantly aids in understanding the test's intent and facilitates collaboration amongst developers. Clear naming conventions make it easy for anyone reviewing your code – yourself included – to quickly grasp the test's purpose, even months later. A consistent and logical structure makes the difference between a confusing mess and a well-organized, easily maintainable test suite.

Avoid tightly coupling your tests to the implementation details. If the internal workings of your function change, your tests should still pass as long as the external behavior remains consistent. This is crucial for maintainability, preventing a cascade of test failures with every minor refactoring. Remember, you are testing the behavior of the code, not the precise how. Instead of checking internal variables directly, assess the function's output against the expected results. This approach creates robust tests that are resilient to implementation changes. Let's say you modify the internal calculation algorithm of your circle area function, the results remain the same, therefore your tests that check the output should pass without needing modification.

Employ mocking effectively when dealing with external dependencies. Mocks simulate the behavior of external components – like databases or network services – allowing you to isolate your unit under test. This prevents tests from failing due to problems outside your code. For instance, if your function interacts with a database, a mock database would return predefined data, ensuring consistent test results regardless of the database's actual state. This isolation ensures that a network outage or a database malfunction will not interrupt your testing process. This is particularly useful when the dependencies are complex or slow, enabling you to run tests more quickly and reliably.

Always aim for high test coverage. While 100% coverage is ideal, it's rarely achievable and sometimes unnecessary. However, striving for high coverage ensures you've tested a significant portion of your code. Tools exist to measure code coverage, showing you which parts of your code are covered by tests and which are not. High coverage is an indicator, not a guarantee, of good quality. Prioritize testing complex or critical sections of the code first. A few well-placed tests in areas with high risk of bugs can often provide a better return on investment than trying to achieve universal coverage.

Write tests early and often. Test-driven development (TDD) encourages writing tests before writing the actual code, forcing you to think critically about the function's requirements. This proactive approach helps uncover design flaws early, reducing the need for extensive refactoring later. Testing early reduces the number of bugs found later in the development cycle, leading to faster turnaround times. It's a preventative rather than a reactive approach.

Leverage a testing framework. Frameworks like JUnit (Java), pytest (Python), or NUnit (.NET) provide structures and tools to simplify the process of writing, running, and managing tests. They offer assertions for verifying expected results, test runners for executing tests, and reporting mechanisms for analyzing test results. These tools reduce manual setup and provide features such as automated test discovery. Don't reinvent the wheel; take advantage of established tools to make testing easier. Choose a framework appropriate for your chosen programming language and project. Regularly review and update your tests. As your code evolves, so should your tests. Outdated tests are ineffective and can provide a false sense of security. Make updating tests a part of your continuous integration and continuous delivery (CI/CD) pipeline, ensuring that tests are always current and aligned with the current code. This continuous feedback loop ensures that the system remains reliable and adaptable in the face of change. It is a good practice to regularly reassess the tests' effectiveness.

2.2 Test-Driven Development (TDD)

Test-driven development, or TDD, flips the traditional software development approach on its head. Instead of writing code first and then tests, TDD prioritizes the creation of tests before writing any functional code. This seemingly counterintuitive method leads to more robust, maintainable, and ultimately, better software. It forces you to think critically about the requirements and design before diving into implementation, resulting in a more clear and focused approach.

This "test-first" approach isn't just about writing tests; it's about using tests as a design tool. The process begins by identifying a specific functionality needed—a single, testable unit. Then, you write a failing test that precisely defines the expected behavior of that unit. This test serves as a concrete specification, ensuring that you fully understand what the code needs to accomplish. Only after this failing test is in place do you write the code to make it pass. Refactoring plays a crucial role in TDD. Once a test passes, you don't just move on. You meticulously review your code, looking for ways to improve its design, clarity, and efficiency. This iterative process of writing a failing test, writing code to make it pass, and then refactoring ensures that your code remains clean, concise, and easy to understand. The tests act as a safety net, allowing you to confidently refactor without fear of breaking existing functionality. It's like having a constantly updated safety check, ensuring the integrity of the existing codebase is maintained.

Consider a simple example: building a function to calculate the area of a rectangle. In TDD, you wouldn't immediately start writing the function. You'd begin with a test that defines the expected behavior: a test that inputs length and width, expecting a specific output (length * width). This test, naturally, would fail initially since no function exists yet. Following this, you develop the function itself, ensuring the test now passes. This ensures that your function correctly carries out its intended purpose, providing a reliable and efficient calculation.

However, TDD isn't a silver bullet. It requires discipline and a shift in mindset. It can feel slower initially, as you're spending time on tests before producing functional code. But this upfront investment pays dividends in the long run, leading to fewer bugs, less debugging time, and a more maintainable codebase. The initial time spent on the tests can save massive amounts of time and effort later on, both in debugging and maintenance.

One common challenge is scope. Writing tests for every tiny detail can be overwhelming. Beginners often struggle with knowing when to stop. The key is to focus on testing the core functionality, the critical aspects that truly determine the success of the unit. Over-testing can be as detrimental as under-testing; finding that balance is a critical aspect of successful TDD. Focus on the main aspects of the function and leave the less important elements for later testing.

Learning TDD takes practice, and patience. Don't be discouraged by initial difficulties. Start with small, manageable units, and gradually increase the complexity. Use a testing framework that is well-suited to your programming language. Many excellent frameworks exist to support the TDD process, providing tools and features that simplify test creation, execution, and reporting.

The advantages of TDD extend far beyond simply catching bugs. The discipline of writing tests beforehand encourages better design, leading to more modular, decoupled code. This modularity makes the software easier to understand, modify, and maintain, reducing long-term costs. This clarity contributes to a more comprehensive understanding of the system and how its components interact.

Finally, remember that TDD is a process, not a rigid methodology. It's iterative and adaptable. There's no single "right" way to implement TDD; rather, it's a set of principles to guide your development process. Experiment, find what works best for you, and don't be afraid to adapt your approach as your projects evolve and your experience grows. It's an iterative procedure of test writing, implementation, and code improvements. Mastering this process greatly enhances the quality and maintainability of your software. The benefits extend to the entire development lifecycle, promoting cleaner, more reliable code.

2.3 Common Unit Testing Pitfalls

Testing too much within a single unit test is a common mistake. Instead of focusing on one specific aspect of a function, developers often try to test multiple functionalities at once. This makes it extremely difficult to pinpoint the exact source of failure should the test fail. A single unit test should focus exclusively on a single unit of code, testing only one specific behavior at a time. This granular approach improves the accuracy and maintainability of your test suite. Isolate each functionality into its separate test method for easier debugging and more reliable results.

Another frequent error lies in neglecting edge cases and boundary conditions. Developers often concentrate on the typical inputs and outputs, overlooking scenarios that lie at the fringes of expected behavior. These edge cases, such as null values, empty strings, or extremely large numbers, can expose hidden flaws in your code. Thorough testing demands careful consideration of these edge cases and boundary conditions to ensure resilience and robustness. Invest time in designing test cases that specifically target these scenarios; it's often where the most unexpected bugs hide.

Failing to use mocking effectively can lead to tightly coupled tests and increased fragility. When tests depend heavily on external systems or databases, small changes in these dependencies can cascade into test failures, even when the tested code remains unchanged. Employing mocking frameworks allows the isolation of units under test from their environment, creating more stable and independent tests that are less susceptible to external influences. By controlling the input and output of dependent components, you obtain a clearer view of your unit's behavior.

Insufficient test coverage is a critical pitfall. Simply having tests is not enough; the tests must adequately cover all branches of code execution paths. Partial coverage leaves significant portions of the code untested, increasing the risk of bugs lurking undetected. Employ code coverage tools to measure the extent of your test coverage and identify areas needing more attention. Strive for high code coverage, understanding that 100% coverage isn't always feasible, but aiming for a high percentage significantly reduces risk. Prioritize testing the most critical and complex parts of your codebase.

Overly complex tests are counterproductive. Tests should be straightforward, easy to read and maintain. If a test is excessively intricate or difficult to understand, it becomes a maintenance nightmare. Refactor overly complex tests into simpler, more focused tests. Maintainability should be a paramount consideration in test design. Clear and concise tests are easier to debug and update as the codebase evolves. Use descriptive names for your test methods to enhance readability and clarity.

Ignoring error handling is a serious oversight. Robust code effectively manages potential errors, and your tests should verify this error handling. Tests should account for exceptional situations, validating how the code responds to unexpected inputs or runtime errors. Always verify that exceptions are handled correctly. Failing to test for error handling can leave vulnerabilities in your code that could lead to unexpected crashes or incorrect results in production environments. Thorough error handling tests build confidence in your code's resilience.

Lastly, neglecting to refactor tests alongside code changes is a common mistake. As code evolves, tests may become outdated or even break. Make sure tests reflect these changes by keeping them up-to-date with evolving code. Tests that consistently fail due to code changes are not serving their purpose. A well-maintained test suite requires consistent attention to ensure tests are current, relevant, and efficient. Keeping tests updated is essential for maintaining a robust and dependable testing regime. Regularly review and refactor your tests to ensure they remain aligned with the current codebase and continue to provide value. Remember, well-maintained tests are an investment in the long-term quality of your software.

3 Integration Testing Strategies

Choosing between top-down and bottom-up integration testing significantly impacts your approach. Top-down testing starts with the highest-level modules, integrating lower-level modules gradually. This method helps early detection of major design flaws, but requires stubbing lower-level modules, potentially masking problems. Bottom-up testing, conversely, starts with the lowest-level modules, progressively integrating higher-level components. This approach thoroughly tests individual modules before integration, but identifying higher-level integration problems might take longer. The optimal choice often depends on the project's complexity and risk profile. A hybrid approach, combining elements of both strategies, can prove highly effective in many scenarios.

Big bang integration, a less commonly favored approach, involves integrating all modules simultaneously and testing the entire system at once. This method is risky, as pinpointing the source of integration problems becomes significantly more challenging. Only consider this approach for smaller, less complex systems with well-understood individual modules. For larger projects, the risk of a chaotic debugging process makes big bang integration far less appealing compared to the controlled approach offered by top-down or bottom-up.

Several integration testing techniques help streamline the process and ensure thoroughness. Incremental integration, for instance, integrates modules one by one, allowing for thorough testing after each addition. This allows for quicker identification and resolution of integration issues. The use of stubs and drivers, while sometimes discussed in the context of unit testing, plays a crucial role here too; stubs simulate the behavior of yet-to-be-integrated modules, while drivers act as stand-ins for calling modules during bottom-up testing. These techniques facilitate testing even when all parts of the system are not yet complete.

Debugging integrated systems presents unique challenges. The complexity of interactions between multiple modules makes identifying the root cause of problems often difficult. Systematic debugging methods are crucial, employing techniques like log analysis, breakpoint debugging, and code instrumentation to pinpoint the origin of errors. Careful use of logging statements at various integration points can provide valuable insights into the system's behavior during testing, making the debugging process more efficient. Remember, a well-structured logging system isn't just for debugging; it provides invaluable information for future maintenance and system monitoring.

Effective integration testing relies on careful planning and a well-defined strategy. Thorough test case design is paramount, ensuring that all interaction points between modules are adequately tested. Test data should be meticulously chosen to cover various scenarios and edge cases. Regression testing, which involves repeatedly running tests after code changes, is crucial for ensuring that new code doesn't introduce new integration issues. Automated testing scripts, when used effectively, can significantly reduce the time and effort involved in regression testing, increasing the overall testing efficiency and reliability.

Remember the importance of utilizing version control. Integrating changes from multiple developers requires careful coordination and tracking of code modifications. Version control systems provide the mechanisms to effectively manage these changes, track bugs and revert to previous versions if needed, allowing for better collaboration and more controlled integration testing processes. This is a crucial aspect of maintainability and system integrity that many overlook, leading to significantly higher levels of frustration and debugging time.

Beyond technical aspects, effective integration testing demands a collaborative approach. Teams must clearly define integration points, communication protocols, and data formats between modules. Good communication between developers and testers is paramount; open channels and clear communication during the testing phase lead to quicker identification and resolution of problems, preventing delays and reducing potential risks. Regular progress reviews help track testing progress, identify and address potential issues early, and ensure the integration process stays on track.

In conclusion, mastering integration testing strategies is essential for delivering high-quality software. The choice between top-down, bottom-up, or hybrid approaches, the selection of appropriate techniques like incremental integration, and the utilization of effective debugging practices are critical factors in a successful integration testing process. However, the human element—collaboration, communication, and systematic planning—cannot be overstated. A successful integration testing strategy isn't just about the tools and techniques; it's about teamwork, discipline, and a commitment to producing reliable software. Remember to prioritize clear communication within your team, leverage effective version control, and use a well-structured logging system to effectively manage and mitigate issues as they arise.

3.1 Top-Down vs. Bottom-Up

Imagine building a house. You could start with the roof, meticulously crafting each shingle before moving down to the walls, then the foundation. That's a top-down approach. Alternatively, you could begin with the foundation, ensuring its stability before constructing the walls, and finally, the roof. This is bottom-up. In software integration testing, these contrasting methods represent significantly different strategies. The choice between them depends heavily on the project's specifics, the team's expertise, and the nature of the software itself.

Top-down integration testing, as the name suggests, starts with the highest-level modules. These modules are tested initially with "stubs"—simplified versions of lower-level modules—that mimic the behavior of the real components. As testing progresses, these stubs are gradually replaced with the actual implemented lower-level modules. This approach allows for early verification of the overall system architecture and major functionalities. It offers the advantage of early detection of major design flaws, potentially saving significant time and resources down the line. However, it requires a deep understanding of the system's high-level design from the start. Furthermore, extensive stub creation can be time-consuming, demanding significant upfront effort and possibly delaying the testing of lower-level functionalities. The risk of overlooking issues in low-level modules remains as those are tested later in the cycle.

Conversely, the bottom-up approach begins with the testing of individual modules—the lowest-level components. These modules are rigorously tested in isolation, ensuring their correctness before integrating them into higher-level modules. As the testing progresses, more and more modules are integrated, building gradually towards a fully integrated system. This method offers a clearer picture of the individual component's functionality. The testing of lower-level modules early in the cycle offers a greater level of confidence in the overall stability and functionality. It tends to be more efficient in detecting and fixing errors at the earliest stage. However, the full system might not be tested until relatively late in the process, potentially uncovering integration issues only once substantial progress has been made. Moreover, the creation of "drivers"—temporary programs to test individual lower-level modules—can add to the overall development complexity.

The choice between these methodologies is not always straightforward. Consider a complex system with many interdependent modules. A top-down approach might reveal early architectural flaws, but could delay the verification of crucial low-level functionalities. On the other hand, a bottom-up approach may detect low-level errors early, but integration problems could emerge late, increasing the risk of costly rework. Often, a hybrid approach, blending aspects of both top-down and bottom-up testing, proves most effective. This might involve a phased integration, starting with testing key high-level modules with stubs initially, then testing crucial low-level components, and later integrating the two. This phased method allows a more adaptive and balanced approach and might be better suited for large projects that involve multiple teams. This approach requires a considerable understanding of the system's architecture and dependencies.

Ultimately, the best approach hinges on factors such as project size, complexity, available resources, and the team's skill set. It's crucial to carefully analyze these factors and determine which strategy or hybrid thereof best suits the specific needs of the project. Regardless of the selected approach, thorough planning, detailed documentation, and efficient communication within the testing team are paramount to ensure the success of the integration testing process. The careful selection of either approach is critical to the outcome of the integration testing and, indeed, the project as a whole. Remember, the ultimate goal is to deliver a high-quality, reliable software product; the best integration testing strategy directly contributes to reaching that goal. It's an essential step in the broader process of rigorous testing and validation to safeguard the integrity of the software.

3.2 Integration Testing Techniques

Big-bang integration testing throws all the modules together at once. It's like building a house by stacking all the materials in a heap and hoping it stands—risky, but potentially fast if it works. The advantage is quick identification of major flaws early on; the disadvantage? Pinpointing the source of a problem in a chaotic jumble of components can be a nightmare. You'll need excellent debugging skills and a solid understanding of the system's architecture to even begin. Thorough logging and monitoring are essential to track down the root cause of failures.

Incremental integration testing is a more measured approach. It's like constructing the house room by room; you test each section before adding the next. This method involves integrating modules in small groups, verifying their interaction before incorporating additional modules. This reduces the complexity and makes fault isolation significantly easier. However, it requires careful planning and a well-defined integration sequence. You might find yourself needing to adjust your testing plans as unexpected problems arise, requiring additional testing on different components.

Top-down integration testing follows a hierarchical approach. Imagine building the house from the roof down: you start with the top-level modules and gradually integrate lower-level modules. This gives you early visibility into the overall system behavior, but it can require the creation of stubs—dummy modules that simulate the behavior of not-yet-integrated components. These stubs can add complexity if not properly handled. Moreover, debugging lower-level modules can be tricky until their higher-level counterparts are integrated.

Bottom-up integration testing is the opposite. It's like starting with the foundation: You begin with the lowest-level modules, integrating them into increasingly larger clusters, and eventually testing the top-level. This method often uses drivers—test programs that simulate the behavior of higher-level modules. The approach provides early verification of the low-level functionality, providing a stable base for integrating higher levels. However, high-level functionality is only tested in later stages, potentially delaying the identification of critical design flaws.

Sandwich testing combines both top-down and bottom-up approaches. Visualize building the house from both ends: You begin with the top level and the bottom level simultaneously, then start integrating the middle layers. This method leverages the strengths of both techniques. It's particularly effective in larger, complex projects where the critical paths can be identified for faster integration testing. The challenge lies in coordinating the different integration pathways and managing the interaction between top and bottom integrations.

Regression testing after each integration step is crucial. It's about ensuring that new integrations haven't broken existing functionality, much like regularly inspecting the house's structure during construction to ensure everything is sound. Automated tests, if used throughout, should make this significantly easier, helping to identify and correct issues rapidly. Without it, even minor integration issues can compound. Always plan for enough time in your schedule for this to avoid project delays.

Choosing the right integration testing technique depends on the project's complexity, dependencies between modules, and available resources. There's no one-size-fits-all solution. Consider the project's architecture, the team's expertise, and the risk tolerance when deciding. It's about finding the balance between thoroughness and efficiency. A well-thought-out strategy can significantly reduce development time and costs while ensuring robust software quality.

Documentation is key. Detailed design documents, specifications, and test plans are essential for both planning and executing integration testing. These documents provide context and guide testers and developers through the process. They provide vital details, reducing the ambiguity of testing. Moreover, well-documented tests act as invaluable references for future maintenance and troubleshooting. Think of them as blueprints of the testing, and the house, ensuring clarity and reducing the risk of misunderstandings.

Selecting the right testing tools is also important. There's a plethora of specialized tools that enhance the integration testing process. These tools can automate tasks like test execution, data generation, and result analysis. The right tools can significantly increase efficiency, reducing manual work and improving accuracy. However, remember that not all tools are created equal—choose them carefully, based on your project's specific needs and resources. A poorly chosen tool could cost more time and resources than you'd save.

Remember, effective integration testing is iterative. Testing is not a one-off activity, but a continuous process that evolves with the development. It's about finding the bugs early to avoid bigger problems. Regular testing reveals problems in the early stages, greatly reducing the cost of fixing them. Moreover, this iterative approach increases confidence in the overall quality of the system. It's about a continuous cycle of testing, fixing and retesting.

Finally, a team of qualified and experienced testers is vital for successful integration testing. Experience makes all the difference between identifying obscure issues and simply reporting superficial faults. Expertise in debugging, fault localization, and familiarity with the specific technologies used is a tremendous asset. Consider the expertise of the testers when planning. The quality of your testing is only as good as the competence of those conducting it. Investing in training and skilled professionals is an investment in a higher-quality product.

3.3 Debugging Integrated Systems

Debugging integrated systems presents unique challenges compared to unit testing. The complexity arises from the interplay of multiple components, each with its own potential failure points. Identifying the source of a bug in this context requires a systematic and methodical approach, often involving more than just examining the code of a single module. Effective strategies leverage logging, debuggers, and a keen understanding of system architecture.

A crucial first step is understanding the system's architecture thoroughly. Knowing the data flow between components is critical. Visual aids, like flowcharts or UML diagrams, can greatly enhance comprehension, especially in complex systems. Without a solid grasp of the system's design, isolating the problem will feel akin to searching for a needle in a haystack. Careful analysis of system specifications and design documents is paramount; this preparatory work saves valuable time in the long run.

Once you've understood the architecture, reproduce the bug consistently. Inconsistencies in the bug's behavior make debugging significantly harder, as it will hinder your ability to isolate the root cause. Develop a detailed set of steps to reliably reproduce the issue. This reproducible test case becomes your primary tool for verifying fixes and ensuring that the bug doesn't reappear unexpectedly. Document every step meticulously; this detailed record is an invaluable asset throughout the debugging process.

Logging is an invaluable tool in this context. Strategically placed log statements can offer significant insight into the system's internal behavior. Include timestamps, relevant variable values, and clear descriptions of each step, especially those close to the point of failure. The goal is to create a detailed audit trail that can illuminate the sequence of events leading to the bug. While verbose logging might seem inefficient, it's often far less costly than hours of unproductive debugging. Prioritize meaningful data – don't log every single variable unless it truly aids diagnosis.

Debuggers are invaluable for stepping through the code, examining variable values, and understanding execution flow. Use breakpoints strategically, focusing on areas where you suspect problems may lie. Set breakpoints in different components to trace the bug's path through the system. Visualizing the system's state at various points offers important clues. Modern debuggers offer advanced features like watchpoints, which trigger breakpoints when specific variables change, allowing a more targeted debugging approach. Don't overlook the power of your debugger; proficient usage vastly accelerates the debugging process.

If the bug still eludes you, consider employing techniques such as binary search. Divide the system into smaller, manageable subsets. By systematically isolating and testing each subset, you can more efficiently narrow down the location of the problem. This divide-and-conquer approach is particularly effective in larger systems, turning a potentially overwhelming task into a series of smaller, more manageable ones. Remember to meticulously document each step in this process.

When all else fails, seek external help. Explain the problem clearly, including the steps to reproduce it, relevant logs, and system architecture information. A fresh perspective can often uncover subtle issues missed during individual debugging efforts. Collaboration with colleagues brings diverse expertise to the table, accelerating problem resolution. Don't hesitate to consult experienced developers, and don't underestimate the power of a second pair of eyes. A well-explained problem often quickly yields a solution with collective brainstorming.

Finally, once the bug is fixed, don't simply move on. Implement robust regression tests to prevent it from reappearing in the future. These tests will verify that future changes to the system don't unintentionally reintroduce the bug. Integrate these tests into your automated testing suite to guarantee long-term system stability and reliability. This proactive approach is far more effective than reacting to the same bug's repeated occurrences. Thorough testing is essential for the long-term health of your software system.

Remember, preventative measures are far more efficient than reactive ones, saving valuable time and resources.

4 System Testing & Validation

System testing is the culmination of your testing efforts, a crucial phase where the entire software system, all its integrated components, is rigorously examined. Think of it as the final dress rehearsal before the grand premiere. This involves not just verifying that individual parts work, but that they work together seamlessly, fulfilling the overall intended functionality. Failing to adequately test the system as a whole can lead to catastrophic failures in production, causing significant financial and reputational damage. Thorough system testing prevents such disasters.

Unlike unit or integration testing, system testing takes a holistic approach. It's not about isolated components; it's about the integrated system, operating in a simulated real-world environment. This requires careful planning and the selection of appropriate testing techniques. Consider factors like the system's intended usage, target audience, and potential vulnerabilities. Each of these aspects will influence your testing strategy.

Functional testing forms a critical aspect of system testing. This involves verifying that the system behaves as specified in its requirements document. Do all features work as intended? Are the user flows intuitive and error-free? Does the system handle all expected inputs correctly, and gracefully handle unexpected ones? These are crucial questions that functional testing aims to answer. The process involves running a suite of tests that cover all functionalities, documenting the results, and analyzing any discrepancies found. Remember to test both positive and negative scenarios, ensuring robustness.

Beyond functional correctness, non-functional requirements must also be tested. This encompasses areas like performance, security, and usability. Performance testing assesses the system's responsiveness, scalability, and stability under various load conditions. Security testing looks for vulnerabilities that might be exploited by malicious actors. Finally, usability testing, often involving real users, ensures the system is intuitive, efficient, and pleasant to use. These aspects are equally critical for a successful software product. A system that functions perfectly but is slow, insecure, or difficult to use will not meet user expectations.

User Acceptance Testing (UAT) is a critical stage in system testing, typically involving real users interacting with the system. Their feedback provides invaluable insights into the system's usability and overall acceptability. It often identifies issues that would be missed by developers or testers. This user feedback is crucial for fine-tuning the system before release, ensuring it aligns with user needs and expectations. Don't underestimate the power of involving the end-user early and often; it's an investment that yields significant returns.

Regression testing is an ongoing process, not just a phase. It's crucial throughout the development lifecycle, especially after code changes or bug fixes. The goal is to ensure that new code hasn't introduced new bugs or broken existing functionality. A well-designed regression testing suite will detect such regressions quickly, minimizing the risk of releasing a flawed system. Automation plays a significant role here; automating regression tests ensures they can be run quickly and frequently, leading to faster feedback and quicker resolution of issues.

Test automation is essential for efficient and scalable system testing. Automated tests can run repeatedly without human intervention, saving time and resources. This enables quicker feedback cycles, allowing developers to address issues more promptly. Automation is particularly effective for repetitive tasks like regression testing, performance testing, and security scans. However, remember that automation is not a replacement for manual testing; it should be used to complement, not replace, human judgment. Careful planning and selection of automation tools are critical for success. Consider factors like the testing framework, integration with existing systems, and ease of maintenance. Remember to balance the investment in automated testing with its benefits. A comprehensive strategy incorporating both manual and automated testing will likely yield the best results.

4.1 Functional and Non-Functional Tests

Functional tests verify that the software behaves as specified in the requirements document. These tests are crucial because they directly address whether the software meets its intended purpose. Imagine a banking application; a functional test might check if a user can successfully deposit funds, transfer money, or view their account balance. These tests are designed to ensure all features work as expected, validating each function against the documented specifications. Failures in functional tests directly indicate deviations from the expected behavior and demand immediate attention. The process involves carefully defining test cases that cover all aspects of the software's functionality. Thorough functional testing is paramount for delivering a reliable and user-friendly product.

Non-functional tests, conversely, evaluate aspects beyond functionality, focusing on attributes like performance, security, and usability. These tests indirectly influence user experience and system stability. For our banking application, a non-functional test might measure the response time for transactions, ensuring it remains within acceptable limits. Another might assess the system's vulnerability to various attacks, checking for security breaches. Usability testing might involve observing how easily users navigate the interface and perform tasks. Non-functional tests are equally vital as functional tests, often determining the overall success of the application, even if all functional requirements are met. A perfectly functional application that is too slow or insecure will ultimately fail to meet user expectations.

Consider the distinction between measuring response times (non-functional) and verifying the accuracy of a transaction (functional). Both are necessary for a robust banking application. While functional tests answer "Does it work?", non-functional tests explore "How well does it work?" and "Is it safe to use?". This interplay between functional and non-functional testing provides a comprehensive evaluation of the software's quality and suitability. A well-rounded testing strategy must incorporate both, creating a complete picture of the system's capabilities and limitations. Omitting either would leave significant gaps in the software's validation process.

The effectiveness of both functional and non-functional tests depends heavily on careful planning and execution. Test cases need to be meticulously designed to cover a wide range of scenarios, accounting for both typical and edge cases. For instance, a functional test for a login system shouldn't only check valid credentials but also handle invalid ones, password resets, and other scenarios. Similarly, non-functional tests need to define specific performance goals and security standards to measure against, ensuring consistency and reliability across testing iterations. Automation can significantly improve efficiency, especially for repetitive tests. Designing effective functional tests often involves using established methodologies like boundary value analysis and equivalence partitioning. Boundary value analysis focuses on testing values at the edges of input ranges, identifying potential errors that might not be apparent with typical input values. Equivalence partitioning divides input data into groups, selecting representative values from each group for testing. These techniques ensure comprehensive coverage without redundant testing. In contrast, non-functional tests might use tools and frameworks specifically designed for performance testing, security analysis, or usability evaluation. Remember, choosing the right tools and techniques is vital for achieving optimal test results.

The results of both functional and non-functional tests are crucial for making informed decisions about software quality and release readiness. Detailed reports must be produced, providing clear summaries of the tests performed, any issues discovered, and their severity. These reports are essential for communicating the state of the software to stakeholders, including developers, testers, project managers, and clients. Thorough documentation is vital not only for immediate decision-making but also for future reference, aiding in debugging, maintenance, and even future development. A robust test report will often include metrics such as pass/fail rates, performance benchmarks, security vulnerabilities identified, and usability scores.

Finally, integrating functional and non-functional tests into a comprehensive continuous integration/continuous delivery (CI/CD) pipeline is increasingly common, ensuring testing is an integrated part of the software development lifecycle. Automation plays a significant role in this integration, enabling frequent and automated tests at different stages of development. This approach helps identify and address issues early, minimizing costs and improving overall software quality. Furthermore, continuous testing promotes faster release cycles without sacrificing quality. Adopting a robust and integrated testing strategy, encompassing both functional and non-functional aspects, is paramount to developing high-quality, reliable software.

4.2 Performance and Security Testing

Performance testing measures how well a software application responds under various workloads. We look at response times, resource utilization (CPU, memory, network), and scalability. For instance, if your e-commerce site experiences a surge in traffic during a holiday sale, will it handle the increased load without crashing or significantly slowing down? Performance testing answers this critical question, identifying bottlenecks and areas needing optimization. Think of it like stress-testing your car engine – you push it to its limits to see how it holds up before facing real-world conditions. Understanding this allows you to build systems capable of handling predicted, and often unpredictable, peaks in usage.

Security testing aims to identify vulnerabilities in an application that could be exploited by malicious actors. This isn't just about preventing unauthorized access; it's about protecting sensitive data and maintaining the integrity of the system. Techniques range from penetration testing (simulating real-world attacks) to static and dynamic code analysis to identify security flaws at the coding level. Consider the potential impact of a successful attack: data breaches, financial losses, reputational damage, and even legal consequences. A robust security testing strategy, therefore, is crucial not only for protecting your users but also for maintaining the trust they place in your software.

Load testing is a subset of performance testing focusing on the application's behavior under high user load. This involves simulating a large number of concurrent users accessing the system, allowing you to gauge how it performs under stress. By monitoring response times, error rates, and resource consumption under various load levels, you can pinpoint the breaking point of your application and identify potential areas for improvement. It's not enough for your software to work perfectly under normal conditions; it must also gracefully handle extreme conditions. This helps anticipate and prevent failures during peak usage periods, guaranteeing user experience remains consistent, even under significant demand.

Stress testing goes a step further than load testing. Here, the goal is to push the system beyond its normal operational limits to identify the point of failure. You might simulate a significantly larger user load than expected or introduce artificial problems (like network failures) to see how the application responds. Stress testing helps uncover vulnerabilities that load testing might miss, providing valuable insights into the system's resilience and robustness. Understanding these breaking points is essential for designing systems that can handle unforeseen circumstances and maintain stability even when faced with extreme challenges. This ensures that when unexpected spikes occur the system can either withstand them or fail gracefully.

Vulnerability scanning is a crucial aspect of security testing. Automated tools can be used to scan the application's code and infrastructure for known security weaknesses. This helps identify potential vulnerabilities, such as SQL injection flaws, cross-site scripting (XSS) vulnerabilities, or insecure authentication mechanisms. While this is not a replacement for more thorough penetration testing, vulnerability scanning is invaluable for rapidly identifying potential risks and prioritizing remediation efforts. This early detection can prevent costly exploits or security breaches and ultimately reduce financial risk.

Penetration testing is a more advanced form of security testing involving a simulated attack on your application to identify vulnerabilities that automated tools might miss. Experienced security professionals attempt to breach the system using various techniques, mimicking real-world attacks, thus identifying weaknesses in security controls. The aim isn't just to find vulnerabilities but to understand how they can be exploited and the potential impact of a successful attack. This provides a realistic assessment of your application's security posture and guides the development of effective mitigation strategies. It also emphasizes prevention rather than reaction.

Security testing also includes examining authentication and authorization mechanisms. This involves testing how users authenticate to the system (e.g., passwords, multi-factor authentication) and whether they only have access to the resources they're authorized to use. Weak authentication mechanisms are a common entry point for attackers, making this a critical area of security testing. This encompasses thorough testing of access controls, verifying that only authorized personnel can access sensitive data or functionality. Strong authentication practices and well-defined access controls are critical to protect your system from unauthorized access and data breaches. Failing to properly manage this could lead to significant issues.

Performance and security testing are not separate but complementary processes. A secure application that performs poorly can still be vulnerable. Conversely, a high-performing application with significant security weaknesses presents a significant risk. A holistic approach integrating both performance and security testing throughout the software development lifecycle is paramount for building reliable and secure applications. Combining these elements ensures both a positive user experience and a secure system that protects valuable assets.

4.3 User Acceptance Testing (UAT)

User acceptance testing, or UAT, is the final crucial step before releasing software. It's where real users, not developers, interact with the finished product, providing invaluable feedback. Think of it as a dress rehearsal before the grand opening night—a chance to identify any remaining issues from the perspective of those who will actually use the system. This ensures the software meets its intended purpose and user expectations. It's not about finding every single bug; it's about validating the system's usability and functionality within the context of its real-world application.

Unlike earlier testing phases focusing on individual components or integrated modules, UAT assesses the system holistically. It's about verifying that all the pieces work together seamlessly to deliver the intended user experience. For example, imagine a new e-commerce platform. Unit tests might verify that the individual shopping cart function works correctly, while integration tests might check that the cart interacts correctly with the payment gateway. UAT, however, focuses on the overall shopping experience—can a user easily browse, add items, proceed to checkout, and complete the purchase? Does the design feel intuitive and user-friendly? These are questions only real users can convincingly answer.

Planning for UAT requires careful consideration. First, identify your target users. These individuals should accurately represent the intended audience for your software. Next, define specific test cases based on typical user workflows and scenarios. The goal is to cover a broad range of functionalities and user interactions. It's advisable to create detailed test scripts outlining each test scenario, including expected outcomes. Providing clear instructions and training to your UAT participants will enhance the effectiveness of the process. These instructions should clearly outline how to use the software and what aspects to test. This will ensure consistent feedback and avoid the need for extensive follow-up.

The execution phase of UAT involves users performing the predefined test cases and documenting their findings. Use a system that allows for easy defect reporting, making it straightforward to log issues with their context. Efficient defect tracking and documentation allow for faster resolutions and provide valuable insights into recurring issues, revealing areas for potential improvement. These reports should not only detail bugs but also include feedback on usability, user experience, and any other relevant aspects of the software's performance. This feedback is more than bug reports; it is invaluable qualitative data, leading to more refined and user-centric products.

Analyzing the UAT results is a critical step that shouldn't be rushed. This involves consolidating feedback from all testers, prioritizing issues based on severity and impact, and working with the development team to address the identified problems. The goal isn't simply to fix the problems; it is to make the software more usable and achieve a more satisfying user experience. This may involve making design adjustments, modifying functionality, or improving the overall user interface. Effective communication throughout this phase is paramount, ensuring that the feedback is clearly understood and acted upon.

UAT is not just about finding and fixing bugs; it's a vital process that validates the product's suitability for its intended users. It builds user confidence by demonstrating a commitment to quality and user satisfaction. This process goes beyond just fixing problems; it improves the user experience, increases user adoption, and ultimately contributes to the success of your software. It's an investment in the long-term usability and success of your product, ensuring that the final product aligns with its intended purpose and satisfies the needs of its end-users. Therefore, a well-executed UAT translates to a more user-friendly, robust, and ultimately more successful product launch. The value of user feedback cannot be overstated; it's the final and essential check before releasing your software to the market. Taking the time and resources to execute thorough UAT will ultimately save you from more significant, costly issues down the road.

4.4 Regression Testing

Regression testing is crucial; it safeguards against unintended consequences. After any code modification, big or small, regression tests ensure existing functionality remains intact. Imagine building a house: each new addition shouldn't compromise the foundation. Similarly, new features shouldn't break existing ones. This prevents costly bugs from slipping into production, causing user frustration and reputational damage. Thorough regression testing acts as a safety net.

Effective regression testing hinges on a well-defined test suite. This suite should comprehensively cover all critical functionalities. The tests themselves need to be robust, reliable, and easily maintainable. Think of this suite as a detailed blueprint, carefully designed to check every aspect of your software. Regularly updating this suite is essential; as your software evolves, so should your tests. This ensures you're always testing the current state of your product.

Automated regression testing is undeniably the most efficient approach. Manually running hundreds or thousands of tests after each minor code change is impractical, and incredibly time-consuming. Automation allows for rapid execution, identifying issues quickly. Tools like Selenium, JUnit, and pytest offer streamlined automated testing capabilities. However, the initial investment in automation requires careful planning and consideration, but the long-term benefits far outweigh the setup costs.

Choosing the right testing approach is pivotal. Selecting the wrong one can lead to unnecessary complexities and inefficient resource allocation. For instance, full regression testing after every small change might be overkill, slowing down the development process. Instead, consider prioritizing tests based on risk. Focus on areas with a higher likelihood of being affected by changes. This targeted approach optimizes efficiency without compromising essential checks.

Prioritization is an art, not just a process. It requires a deep understanding of the software architecture and the impact of various code modifications. Experience plays a significant role here; identifying potential hotspots becomes intuitive with practice. Effective prioritization is a skill honed through years of experience, but beginners can benefit from observing experienced testers and using code analysis tools. These tools often highlight crucial aspects of the code, allowing them to zero in on potential problem areas.

The selection of appropriate test data is equally important. Data must accurately reflect real-world scenarios; otherwise, the test may miss critical issues. Consider using various datasets and input parameters to ensure robustness. For example, boundary testing can reveal flaws by checking how the software behaves at the extremes of its input range. Remember, thorough test data generation is as important as the tests themselves. The selection of appropriate data impacts the reliability of your results, therefore impacting the whole regression test success.

Documentation is the unsung hero of regression testing. Well-documented tests are easier to understand, maintain, and debug. Clear descriptions of test cases, expected results, and steps ensure consistency and collaboration among team members. This also aids future maintenance and troubleshooting, saving considerable time and effort in the long run. Without it, your tests become a puzzle, difficult to resolve.

Integrating regression testing into your development lifecycle is essential. Continuous integration and continuous delivery (CI/CD) pipelines automate regression testing as part of the build process. This ensures that any code changes are immediately tested, preventing major issues from creeping into later stages. Early and regular testing reduces the risks and costs associated with fixing bugs later in the cycle. It's also important to have a system where reporting is clear and well-organized, simplifying the identification and resolution of any regressions.

Analyzing the results of your regression testing is the final, yet critical, step. Don't just blindly accept the results. Investigate failures carefully, understanding their root cause, not merely the symptoms. This analysis is essential for improving both your software and your testing strategy. It helps identify areas that need further attention and enhances the effectiveness of your testing methodology. Regular review of these results and changes of approach is what makes regression testing a continuous improvement loop.

Finally, regression testing isn't merely a checklist; it's a continuous process of learning and refinement. Each test cycle provides valuable feedback, improving your understanding of the software and identifying areas for improvement. Consider it a journey of continuous improvement, not merely a destination to be reached.

Every iteration brings new knowledge, which should translate into better tests, and ultimately, better software. The long-term benefits of a robust, well-maintained regression testing strategy far outweigh the initial effort.

4.5 Test Automation

Automating your testing process is crucial for efficiently scaling your software development lifecycle. Manual testing, while valuable in certain contexts, becomes increasingly impractical and prone to human error as the complexity and size of your project grows. Automation allows for repetitive testing, ensuring consistency and reducing the likelihood of overlooking critical issues. This translates directly to cost savings and improved time-to-market. Think of it as building a tireless, highly accurate, digital QA team that works 24/7.

Selecting the right automation framework is paramount to success. Your choice will heavily depend on factors like programming language proficiency within your team, the existing tech stack of your project, and the specific type of testing you need to automate (unit, integration, system, etc.). Popular frameworks like Selenium for web applications, Appium for mobile apps, or JUnit/TestNG for Java-based projects offer robust capabilities. Don't rush into a decision; thorough research and potentially even pilot testing different options are vital before committing to a particular framework. Ignoring this crucial step can lead to long-term compatibility issues and developer frustration.

The design of your automated tests is critical to their effectiveness. Avoid hardcoding values whenever possible; utilize data-driven testing methodologies to run the same test with multiple inputs. This approach significantly increases test coverage and enhances the robustness of your automation suite. Imagine a single automated test capable of handling thousands of variations simply by modifying the input data. Furthermore, ensure your tests are modular and independent; a failure in one test shouldn't cascade and compromise the entire testing process. Strive for clear and concise test scripts, making them easily readable and maintainable for your team. Poorly designed tests become a nightmare to debug and update, negating the very benefits of automation.

Implementing continuous integration (CI) is essential when incorporating test automation into your workflow. CI tools, such as Jenkins, GitLab CI, or Azure DevOps, automate the process of building, testing, and deploying your software. By integrating your automated tests into your CI pipeline, you create a feedback loop that catches potential issues early in the development process. This drastically reduces the cost of fixing bugs later on and allows for quicker iterations. Essentially, each code commit triggers an automated build and test cycle, alerting developers to any failures immediately. Failing to integrate your tests into a CI/CD pipeline significantly weakens your automation efforts, limiting its full potential to enhance your development lifecycle.

Maintaining your automated tests is an ongoing effort, not a one-time task. As your software evolves, your tests must adapt. Prioritize writing maintainable and easily understandable tests. This reduces the maintenance burden in the long run and makes it simpler for future developers to modify and enhance your existing test suite. Regularly review your test suite to identify any obsolete or redundant tests. Removing outdated tests improves the efficiency and effectiveness of your test automation strategy. Consider employing code reviews and establishing best practices within your team to ensure consistent quality and maintainability. A neglected test suite quickly becomes a liability, eventually outweighing the benefits of automation. Proper maintenance is fundamental to the longevity and success of your automated testing strategy.

Finally, remember that test automation is not a replacement for manual testing. While it automates the repetitive aspects, it cannot fully replace the human element, especially in areas that require nuanced judgment, exploratory testing, and user experience evaluation. A balanced approach that combines both automated and manual testing provides the most comprehensive and robust quality assurance strategy. Automation is a powerful tool, but it's not a magic bullet. Understanding its limitations and strategically integrating it within your overall testing plan will yield the most substantial and sustainable improvements in your software development process.

5 Advanced Testing Techniques

Static analysis scrutinizes code without execution, identifying potential vulnerabilities and bugs early in the development cycle. Think of it as a thorough code review performed by a highly analytical tool, flagging suspicious patterns, style violations, and potential security risks. This proactive approach can save considerable time and resources later, preventing costly fixes during later stages. Tools like Lint and FindBugs are prime examples, adept at identifying issues ranging from simple style inconsistencies to more complex memory leaks. Integrating static analysis into your continuous integration pipeline is a best practice, ensuring that potential problems are detected before they reach production. This early intervention significantly reduces the risk of deploying flawed software.

Dynamic analysis, in contrast, involves running the software and observing its behavior under various conditions. It's a more hands-on approach, providing real-time insights into how the code executes and interacts with its environment. Dynamic analysis techniques include runtime testing, memory debugging, and performance profiling. Imagine it as a detective examining a crime scene, carefully observing and analyzing the evidence (the software's behavior) to uncover any anomalies. Tools like Valgrind and gdb provide deep insights into memory allocation, program flow, and performance bottlenecks. Combined with static analysis, dynamic analysis gives you a comprehensive view of your software's behavior and potential weaknesses. Remember that while static analysis can prevent many issues, dynamic analysis is essential for finding bugs that only surface during runtime.

Code coverage analysis quantifies the extent to which your test suite exercises different parts of your codebase. It's a crucial metric for determining the thoroughness of your testing efforts. A high code coverage percentage, say 80% or more, indicates that a large portion of your code has been tested. However, it's important to remember that high code coverage doesn't guarantee perfect software. Even with high coverage, poorly designed tests might miss crucial bugs. Tools like JaCoCo and Clover generate code coverage reports, visually highlighting the parts of your code that have been tested and those that haven't. Using this data, you can strategically focus your testing efforts on the untested parts, maximizing the effectiveness of your test suite and improving overall software quality. Regular code coverage analysis is an indispensable part of modern software development.

Mutation testing is a more advanced technique that goes beyond simply executing tests. It involves systematically introducing small changes (mutations) into your codebase and then running your existing tests to see if they detect those changes. If a mutation goes undetected, it indicates a weakness in your tests. This technique helps identify blind spots in your test suite, revealing areas where additional testing is needed. Think of it as a rigorous stress test for your tests themselves. Mutation testing tools can be computationally expensive, but the insights they provide about the robustness of your test suite can be invaluable in the long run. By addressing the weaknesses revealed through mutation testing, you can dramatically improve the quality and reliability of your software.

Consider implementing fault injection testing to simulate real-world scenarios that might cause software failures. This proactive approach involves intentionally introducing errors, such as incorrect inputs or network interruptions, to assess the software's resilience and error-handling capabilities. It's like conducting a controlled experiment, observing the software's reaction to different forms of stress. The software's response to these injected faults can reveal hidden vulnerabilities and improve fault tolerance, which can be crucial in critical systems. Fault injection testing helps identify vulnerabilities before they're exploited in production. By understanding how your software behaves under stress, you can build more robust and reliable systems.

Finally, remember that advanced testing isn't just about using sophisticated tools. It's also about adopting a robust testing mindset. Thoroughly review code, design comprehensive test plans, and continually refine your testing strategies. Continuously evaluating and improving your testing processes is a commitment to building high-quality software that meets the needs of your users and remains resilient against unexpected challenges. The value of rigorous, consistent testing should be recognized, not as an additional task, but as an integral and indispensable part of the software development lifecycle. It's an investment in quality and reliability that pays dividends in the long run.

5.1 Static and Dynamic Analysis

Static analysis scrutinizes your code without actually running it. Think of it as a meticulous proofreader, examining every line for potential problems before the program even starts. This involves tools that dissect the code's structure, hunting for vulnerabilities like buffer overflows, potential null pointer dereferences, or even subtle logical errors that might slip past a human's eye. The benefit? Early detection of problems saves time and resources in the long run, preventing costly fixes later in the development cycle. Ignoring static analysis is akin to building a house without blueprints; you might get lucky, but the odds of structural flaws are significantly increased.

Dynamic analysis, conversely, involves executing your code and observing its behavior in a controlled environment. Imagine it as a detective, meticulously tracing the code's execution path, noting its response to various inputs, and looking for unexpected behaviors. This method can uncover runtime errors, memory leaks, and performance bottlenecks that static analysis might miss. Dynamic analysis often incorporates techniques like fuzzing, which bombards the software with random or unusual input, to reveal unexpected vulnerabilities. It's a powerful tool, but it's not without its limitations. It requires extensive test cases and comprehensive coverage to maximize its effectiveness. Finding subtle bugs often requires a deep understanding of the codebase.

Choosing between static and dynamic analysis isn't an either/or proposition. A robust testing strategy intelligently combines both approaches, leveraging their complementary strengths. Static analysis acts as the first line of defense, catching many common errors early. Dynamic analysis then refines this process, providing a deeper understanding of runtime behavior and uncovering issues not readily apparent through static examination. The combination helps provide a holistic view of code quality and security.

Consider a scenario where static analysis flagged a potential null pointer exception. This isn't necessarily a critical error; it could be a false positive. Dynamic analysis steps in, simulating scenarios involving null values. If the null pointer indeed causes a crash, you've confirmed the static analysis finding and identified a critical vulnerability. However, if the program handles the null value gracefully, the static analysis warning can be reassessed and potentially dismissed as a harmless anomaly. This iterative process of verification and refinement significantly enhances the reliability and robustness of your software.

Integrating these methods into your workflow can significantly impact your software's quality and security posture. Tools exist to automate both static and dynamic analyses, seamlessly incorporating them into your development pipeline. This integration allows for consistent testing throughout the development lifecycle. The results from both methods will aid you in promptly identifying vulnerabilities, paving the way for rapid remediation and ultimately improving the long-term maintainability of your software. Moreover, prioritizing both static and dynamic analysis demonstrates a dedication to software excellence, benefiting both developers and end-users alike. This commitment to rigorous testing underscores your professional dedication.

The synergistic use of static and dynamic analysis methods minimizes risk and fosters a proactive approach to error prevention.

By identifying and addressing issues early, you avoid costly rework and minimize the risk of deploying faulty software. Regularly incorporating these practices also contributes to improved developer skills and fosters a culture of quality within your team. This continuous improvement cycle directly translates into more robust and reliable software.

Furthermore, mastering these techniques isn't just about technical proficiency; it's about understanding how they contribute to a broader strategy of comprehensive software quality assurance. Successfully implementing these methods builds confidence in your software's reliability and greatly reduces the likelihood of post-release bugs and security breaches. Remember, these are not isolated practices; they are vital components of a larger strategy geared towards delivering superior, secure software. By combining both, you are creating a stronger defense against potential risks.

Finally, continuous learning is crucial in this ever-evolving field. Staying updated on the latest tools, techniques, and best practices for both static and dynamic analysis is essential for keeping your software at the forefront of quality and security. Attend workshops, read industry publications, and engage with the wider software development community to benefit from shared knowledge and collective experiences. Remember, the pursuit of software excellence is a continuous journey.

5.2 Code Coverage Analysis

Code coverage analysis provides a crucial metric for evaluating the thoroughness of your testing efforts. It quantifies the percentage of your codebase executed during testing, highlighting untested sections that might contain bugs. Understanding this metric is essential for improving software quality and reducing the risk of undetected defects reaching production. This isn't simply about achieving a high percentage; it's about strategically targeting those parts of your code that are most likely to contain errors.

Consider a scenario: You've meticulously tested the happy path, the most common use cases of your application. However, code coverage analysis might reveal that error-handling routines, edge cases, and rarely-used features remain largely untested. These untested areas are often where subtle, hard-to-find bugs lurk. A high overall code coverage doesn't guarantee bug-free software, but significantly lower coverage is a strong indicator of potential problems.

Different types of code coverage exist, each offering a distinct perspective on test thoroughness. Statement coverage, for example, measures whether each line of code has been executed at least once. Branch coverage goes a step further, examining whether each decision point (if/else statements, switch cases) has been tested for all possible outcomes. Path coverage aims to traverse every possible execution path through the code. The choice of coverage type depends on the project's risk profile and the level of assurance required. For critical systems, a higher level of coverage like branch or path coverage is warranted. For less critical applications, statement coverage might be sufficient.

While striving for high code coverage is beneficial, it's crucial to remember that it's not the only indicator of good testing. Comprehensive testing necessitates a well-designed test suite encompassing various test types, including unit, integration, and system tests. Complete code coverage is, therefore, a goal to strive for but not an absolute requirement for quality assurance. Focusing solely on maximizing code coverage while ignoring other critical aspects of testing can lead to an incomplete and inefficient approach.

Using code coverage tools, generally integrated into your development environment or available as standalone applications, is paramount. These tools instrument your code, monitoring execution during tests. They then generate reports visualizing the coverage achieved, precisely identifying untested parts. Many tools provide detailed reports showing exactly which lines or branches of code remain untested. This granular information helps developers pinpoint areas needing additional tests. Popular examples include SonarQube, JaCoCo, and Coverity, each with its own strengths and integration capabilities, offering various features tailored to different languages and environments.

Remember, the goal isn't simply to inflate your code coverage numbers through superficial tests. Instead, aim to develop targeted tests that thoroughly exercise critical sections, including error handling, edge cases, and less frequently used functionalities. A well-designed test suite targeting these specific areas will often provide greater value than a high overall code coverage achieved through less meaningful tests. Analyze the results of your code coverage analysis strategically, identifying which areas need the most attention and prioritizing your testing efforts accordingly.

Integrating code coverage analysis into your development workflow provides continuous feedback on testing effectiveness. Regular analysis, especially during continuous integration/continuous delivery (CI/CD) pipelines, enables early detection of gaps in testing. This proactive approach allows for quicker identification and resolution of potential issues, preventing defects from accumulating and significantly reducing the cost of fixing them later in the development cycle.

Finally, code coverage analysis is an invaluable tool for enhancing your software development process. By combining a strategic approach to testing with the insights provided by code coverage tools, you gain a quantifiable measure of your testing thoroughness. This holistic approach helps build more robust and reliable software applications, leading to higher-quality products and a more efficient development process. Don't view code coverage as a mere metric; view it as a guide towards building a better product, identifying areas for improvement, and ultimately, increasing confidence in your software's reliability.

6 Software Validation Best Practices

Software validation isn't just about finding bugs; it's about building confidence. A robust validation process ensures your software not only functions as intended but also meets its specified requirements, complies with relevant standards, and ultimately, delivers value to its users. This involves far more than simply running a few tests; it's a systematic approach that integrates throughout the entire software development lifecycle.

Formal verification methods offer a powerful, mathematically rigorous approach to validation. Techniques like model checking and theorem proving allow for the automated verification of software properties, eliminating the guesswork inherent in traditional testing. While these methods can be computationally intensive and require specialized skills, their ability to guarantee the absence of certain classes of errors makes them invaluable for safety-critical systems, such as medical devices or aerospace software.

Their application, however, requires a careful consideration of the tradeoffs between rigorous validation and project time constraints.

Software Quality Assurance (SQA) provides a broader framework encompassing the entire software development process. It's not just about testing; it's about establishing a culture of quality from the outset. A robust SQA program includes well-defined processes, clear roles and responsibilities, and a commitment to continuous improvement. Regular audits, code reviews, and adherence to coding standards are vital components, all ensuring the software's quality aligns with its intended purpose. Remember, early detection of issues is significantly cheaper and easier to resolve than dealing with them later in the development cycle, a key aspect to remember when planning your SQA program.

Thorough documentation and traceability are essential for successful software validation. Clear requirements specifications, design documents, test plans, and test results provide a complete audit trail, demonstrating that the software meets its specified requirements. Traceability links each requirement to the corresponding design elements, code segments, and test cases, allowing for easy identification of the root cause of any issues. This provides confidence in the software's integrity and facilitates future maintenance and updates, a crucial aspect for long-term software projects.

Effective validation also necessitates the adoption of a risk-based approach. Identifying potential risks early allows for the prioritization of testing efforts, focusing on the areas most likely to impact the system's reliability and safety. This involves a careful analysis of the software's functionality, its intended users, and its operational environment. A risk assessment will guide the validation process, ensuring that resources are focused where they are needed most. Using a risk-based approach reduces the overall effort while maximizing the effectiveness of testing.

Consider integrating static analysis tools into your validation process. These tools examine your code without actually running it, identifying potential errors such as memory leaks, buffer overflows, and concurrency issues early in development. Static analysis can significantly reduce the number of bugs that make it into later stages, saving both time and money. Remember to combine this with dynamic analysis methods, which involve running the software with various inputs to detect runtime errors. The combination of these two methods provides a comprehensive approach to validation.

Remember, the validation process is not a one-time event but rather an ongoing activity. Regular testing and monitoring throughout the software's lifecycle, especially after updates and changes, are essential for maintaining its quality and reliability. This includes conducting regression tests to ensure that new features or bug fixes do not introduce unexpected problems. Automation plays a significant role in enabling continuous validation, minimizing the time and resources required to conduct this ongoing process.

Finally, remember that the success of software validation heavily relies on a collaborative effort among developers, testers, and stakeholders. Open communication, clear responsibilities, and a shared commitment to quality are crucial factors in building a reliable and successful software system. Regular reviews, feedback sessions, and efficient communication channels help maintain a consistent and unified validation effort. The goal is not just to meet requirements but to exceed expectations, creating software that not only functions correctly but also provides a valuable and user-friendly experience. This holistic, collaborative approach brings a level of quality assurance unmatched by isolated testing efforts.

6.1 Formal Verification Methods

Formal verification methods offer a rigorous approach to software validation, going beyond traditional testing techniques. Unlike testing, which focuses on observing software behavior through execution, formal verification mathematically proves the correctness of a program's logic against its specifications. This involves expressing both the program and its intended behavior using formal mathematical languages, allowing for automated analysis and verification. The benefit? A significantly higher degree of confidence in the software's reliability and safety, especially critical in safety-critical systems like air traffic control or medical devices.

Model checking is a prominent formal verification technique. This method systematically explores all possible states and transitions of a finite-state system, checking whether the system satisfies a given property expressed as a temporal logic formula. Tools like SPIN and NuSMV automate this process, identifying potential errors or proving the absence of specific faults. Imagine a complex state machine controlling a robotic arm; model checking can verify that the arm never enters a state where it collides with its environment, irrespective of the input sequence. This eliminates the possibility of runtime errors discovered only through exhaustive testing.

Theorem proving offers a different approach, using a more general-purpose mathematical framework. Here, you express program properties as theorems and use automated or interactive theorem provers to deduce their validity. This technique is especially powerful for proving complex properties about infinite-state systems, where model checking may fall short. While more demanding in terms of expertise and time investment, theorem proving provides a higher level of assurance, effectively guaranteeing the correctness of certain aspects of a program's behavior. Consider proving the correctness of a sorting algorithm; theorem proving allows a formal demonstration that the algorithm always produces a sorted output for any given input.

While formal methods boast significant advantages, their application presents challenges. The process of modeling a system formally requires specialized skills and can be time-consuming. The complexity of the system can quickly overwhelm the capabilities of verification tools. Furthermore, not all aspects of software behavior are easily expressible in formal languages. The level of abstraction required for formal verification might mean overlooking low-level details that could still lead to errors. For instance, while formal verification might prove the correctness of an algorithm's core logic, it might not catch subtle timing issues or unexpected hardware interactions.

Despite these limitations, formal verification remains an invaluable tool in the software validation arsenal. Its strength lies in providing a level of confidence unmatched by traditional testing methods. It is especially beneficial for safety-critical systems where even a single error can have catastrophic consequences. The increasing sophistication of formal verification tools and a growing body of expertise are gradually lowering the entry barrier, broadening the accessibility of these powerful techniques. Selective application of formal methods to the most critical components of a system, rather than a blanket adoption for everything, offers a practical and cost-effective strategy. Integrating formal verification into a broader testing strategy provides a robust defense against software vulnerabilities, reinforcing confidence in the reliability and correctness of developed software. The future of software engineering increasingly depends on the synergy of rigorous testing techniques, coupled with the strength of formal methods to ensure optimal quality and safety assurance. Strategic application, focusing on the most critical components, ensures efficient and effective use of this technology.

6.2 Software Quality Assurance (SQA)

Software Quality Assurance (SQA) isn't just about finding bugs; it's about building a culture of quality. It's a proactive approach, woven into every stage of the software development lifecycle, ensuring that the final product meets not only its functional requirements but also its broader quality goals. This isn't a separate phase tacked onto the end; it's an integral part of the entire process. Think of it as a constant guardian, watching over the integrity of your software from conception to deployment.

Effective SQA starts with clearly defined requirements. Ambiguity is the enemy of quality. Precisely documented requirements, agreed upon by all stakeholders, create a baseline against which the software's performance can be rigorously measured. This involves more than simply listing features; it's about specifying the system's behavior under various conditions, including edge cases and potential failures. Vague requirements lead to misunderstandings, ultimately resulting in a product that doesn't fully meet the needs of its users. Consider using techniques like user stories or use cases to articulate requirements comprehensively.

Next, establish a robust testing strategy. This strategy needs to incorporate various testing methodologies, such as unit testing, integration testing, system testing, and user acceptance testing (UAT), each serving a different but equally crucial purpose in uncovering potential problems. Don't rely solely on one type of testing; a multi-faceted approach is essential for thorough quality assurance. Each methodology's effectiveness hinges on carefully selected test cases that aim to cover a wide array of scenarios. Adequate planning and resource allocation are paramount for success, including sufficient time for comprehensive testing.

Beyond the technical aspects, SQA encompasses processes and procedures. This involves implementing a rigorous review process, ensuring thorough code reviews and documentation checks. Regular audits of the SQA processes themselves help to identify areas for improvement and maintain consistency in quality. This creates a feedback loop; identifying bottlenecks, improving processes, and streamlining workflows. Documenting every step, from requirement gathering to deployment, provides traceability, making it easy to track changes and the reasons behind those changes. This rigorous documentation also allows for efficient troubleshooting if problems arise.

A critical aspect often overlooked is the human element of SQA. It's not just about the tools and techniques; it's about fostering a culture of quality within the development team. Empower developers to take ownership of their code's quality. Promote collaboration and open communication, creating an environment where identifying and resolving defects is valued and rewarded. Encourage continuous learning and improvement by attending workshops or online training on advanced quality control strategies and best practices. In short, embedding quality is a continuous journey, not a one-time effort.

Furthermore, metrics are essential for monitoring the effectiveness of your SQA efforts. Track key indicators such as defect density, testing efficiency, and customer satisfaction. Analyze these metrics regularly to identify trends and areas for improvement. Data-driven decision-making ensures that your SQA processes are continuously evolving and adapting to the challenges of software development. These metrics shouldn't be used punitively; they are valuable tools for learning and improvement.

Finally, don't underestimate the importance of continuous integration and continuous delivery (CI/CD) pipelines. Integrating automated testing into your CI/CD pipeline allows for continuous monitoring of code quality. This enables quick detection of bugs and minimizes the time it takes to resolve them. Frequent integration and automated tests contribute to a faster and more reliable release cycle. Using tools that facilitate CI/CD processes can significantly improve the efficiency and effectiveness of your SQA activities. This approach directly supports the goal of delivering high-quality software rapidly.

In essence, Software Quality Assurance is a multifaceted discipline that requires a holistic approach. It's about more than just finding bugs; it's about proactively preventing them. By integrating these strategies and fostering a culture of quality, you significantly improve your chances of delivering robust, reliable, and high-quality software that meets user expectations and withstands the test of time. Remember, it's an ongoing process of learning, improvement, and adaptation. The investment in robust SQA practices pays dividends in reduced costs, improved user satisfaction, and a significantly enhanced reputation for producing quality software.

6.3 Documentation and Traceability

Maintaining comprehensive documentation is paramount for successful software validation. Imagine a sprawling city; without detailed maps and records, navigating, making repairs, or even understanding its layout would be impossible. Similarly, without meticulous documentation, your software becomes a confusing maze, difficult to understand, modify, or debug. This documentation serves as a roadmap for developers, testers, and future maintainers, providing critical context for understanding the software's design, functionality, and behavior. Thorough documentation prevents costly rework and ensures consistent quality throughout the software lifecycle.

Effective documentation necessitates a clear and consistent structure, easily accessible to all stakeholders. Think of it like a well-organized library, where each book (module or component) has a clear label (description) and a readily available index (search function). This structure should include detailed descriptions of each module, outlining its purpose, functionality, and interface specifications. Flowcharts, diagrams, and other visual aids can greatly improve understanding, making complex interactions easier to grasp. Furthermore, the documentation must remain consistent throughout the project, avoiding any ambiguity or contradictory information that would only lead to further complications down the line.

Traceability is the crucial link between requirements, design, code, and test cases. It establishes a clear chain of evidence showing how each requirement translates into design elements, which are then implemented in the code, and finally validated through testing. Consider it like a detective's case file, meticulously linking each piece of evidence to form a coherent narrative. This complete traceability ensures that every requirement is accounted for, preventing crucial functionalities from being overlooked or improperly implemented. It is essential for demonstrating compliance with standards and regulations. Moreover, if defects are found, traceability provides a crucial path for isolating the root cause, leading to more efficient and effective debugging.

Implementing traceability requires a structured approach. One effective strategy is the use of unique identifiers for every requirement, design element, code segment, and test case. These identifiers create a direct link, enabling easy tracking and verification. This approach is essential in large and complex projects. Tools such as requirements management systems, version control systems, and test management systems can automate many aspects of traceability, reducing the risk of human error and ensuring complete coverage. Remember, however, that technology is a tool; a sound methodology underpinned by a commitment to meticulous record-keeping remains the true cornerstone of successful traceability.

Beyond the technical implementation, establishing a culture of documentation is essential. This involves promoting a mindset where documentation is seen not as an afterthought, but as an integral part of the software development process. This shared understanding encourages developers to write clear, concise code, accompanied by comprehensive comments and documentation. Regular reviews and updates to the documentation prevent it from becoming outdated or irrelevant. This approach fosters collaboration and reduces the risk of inconsistencies, ensuring everyone works from a common understanding of the system. Moreover, fostering a clear and collaborative approach streamlines project communication and overall performance. This shared ownership results in higher quality software and reduces long-term maintenance costs.

The effort invested in comprehensive documentation and traceability may seem significant upfront, yet the long-term benefits far outweigh the initial cost. Consider the time and resources wasted debugging obscure defects that could have been easily identified with proper traceability. Think of the hours spent reconstructing lost knowledge from an incomplete and unorganized documentation set. The improved maintainability, reduced debugging times, and easier adaptation to future changes resulting from comprehensive documentation and traceability are undeniably substantial. They contribute to building robust, reliable, and scalable software, significantly enhancing its overall quality and longevity. The result is a more efficient and less frustrating development process leading to a better overall product.

7 Conclusion: Securing the Future of Software

The software landscape is constantly shifting, a dynamic ecosystem fueled by innovation and ever-increasing complexity. New technologies emerge daily, demanding equally innovative approaches to testing and validation. Therefore, embracing agile methodologies and adopting a mindset of continuous improvement is no longer optional; it's essential for survival. Adaptability and a willingness to learn are the hallmarks of success in this field, requiring professionals to continuously update their skillsets and knowledge to keep pace with the latest advancements. This ongoing pursuit of excellence isn't just about staying relevant; it's about securing the integrity and trustworthiness of the software we build.

Continuous Integration/Continuous Delivery (CI/CD) pipelines have revolutionized the software development lifecycle. These automated systems streamline the process, integrating code changes frequently and automating the build, test, and deployment phases. This not only accelerates the development cycle but also significantly improves software quality by identifying and resolving issues early on. Think of it as a preventative healthcare approach for your software – regular check-ups and proactive measures minimize the risk of major problems later. Implementing a robust CI/CD pipeline requires careful planning and coordination, but the long-term benefits, in terms of efficiency, reduced costs, and improved product quality, far outweigh the initial investment.

However, technological advancement alone isn't sufficient to guarantee the future of software. Ethical considerations must be at the forefront of every decision, from design to deployment. We, as software developers and testers, bear a significant responsibility to create systems that are not only functional and efficient but also safe, equitable, and privacy-respecting. This includes carefully considering the potential societal impact of our work, addressing issues of bias in algorithms, and protecting user data with the utmost diligence. Ignoring these ethical implications is not just irresponsible; it can lead to significant harm, undermining the very trust we strive to build. The creation of secure and responsible software systems should be an integral part of our professional code of ethics.

Software testing is evolving beyond its traditional role. It is becoming more integrated into the entire software development process, encompassing not just bug detection, but also preventative measures to minimize issues early on. This requires collaboration between developers, testers, and stakeholders, creating a culture of shared responsibility for software quality. We need to proactively identify potential risks and vulnerabilities throughout the development process, incorporating security testing as an integral part of every stage. The shift is from simply testing for defects to ensuring the security and ethical integrity of software from inception to deployment. It's no longer enough to build something functional; we must build something responsible and secure.

In conclusion, securing the future of software demands a multifaceted approach. It requires continuous learning, embracing innovative technologies like CI/CD, and prioritizing ethical considerations in every step of the development process. It's about fostering a culture of responsibility and collaboration, where developers and testers work hand-in-hand to build secure, reliable, and ethically sound software systems. By prioritizing quality, security, and ethical considerations, we build trust, ensuring the long-term sustainability and positive societal impact of software technology. This proactive and holistic approach is crucial not only for the continued success of our industry, but also for the well-being of users and society as a whole. The future of software depends on our commitment to excellence, ethical behavior and proactive risk management. Ultimately, it is about using our expertise and skill to create a technological landscape that benefits all of humankind. Only through careful consideration and ongoing adaptation can we guarantee the future of software will be both innovative and responsible.

7.1 The Evolving Landscape of Testing

The software testing field isn't static; it's a dynamic landscape constantly evolving to meet the demands of increasingly complex software systems. New technologies, methodologies, and challenges emerge regularly, requiring testers to adapt and refine their approaches. Consider the rise of cloud-native applications and microservices architectures: these distributed systems present unique integration and testing challenges that traditional methods might not adequately address. The sheer scale and complexity necessitate innovative testing strategies focusing on resilience, scalability, and fault tolerance. This requires a shift towards automated testing, leveraging tools designed for continuous integration and continuous delivery (CI/CD) pipelines.
Furthermore, the increasing reliance on Artificial Intelligence (AI) and Machine Learning (ML) within software applications introduces another layer of complexity. Testing AI systems requires a

different approach entirely, moving beyond traditional functional testing to encompass assessment of the model's accuracy, bias, and robustness in real-world scenarios. Techniques such as adversarial testing and explainable AI (XAI) are becoming crucial in building trust and ensuring responsible AI development. These advanced testing methods need specialized expertise, highlighting the need for ongoing professional development and upskilling within the testing community.

The shift towards agile and DevOps methodologies also significantly impacts testing practices. The rapid iteration cycles characteristic of these approaches demand integrated testing strategies incorporated throughout the software development lifecycle. This requires a collaborative approach between developers and testers, fostering a shared responsibility for quality. Continuous testing, integrated within the CI/CD pipeline, enables early detection of defects, minimizing the cost of remediation. This necessitates a move towards automation, and a shift from merely identifying bugs, to preventing them entirely through robust coding practices and continuous monitoring.

Beyond technical advancements, the evolving landscape of testing also reflects an increasing emphasis on security. Software security is no longer a separate concern but an integral part of the development process. Secure coding practices must be implemented from the outset, and rigorous security testing is essential at every stage. This includes penetration testing, vulnerability assessments, and security audits to identify and mitigate potential vulnerabilities before deployment. The rise of sophisticated cyber threats demands proactive and responsive security testing approaches.

Finally, the evolving landscape of testing is inextricably linked to the ethical considerations surrounding software development. Testers have a responsibility to ensure that software is not only functional and secure but also ethical and unbiased. This includes assessing potential societal impacts, considering the implications of AI systems, and ensuring that software does not perpetuate harmful biases or contribute to social inequalities. A commitment to responsible software development is crucial, and testing plays a vital role in achieving this goal. The future of software testing involves not just technical proficiency but also a commitment to ethical principles and social responsibility. As software continues to shape our world, ethical testing practices will become even more critical. The responsible development and deployment of software are paramount, and the role of software testing is to ensure that these principles are upheld and maintained in a constantly changing technological world. This responsibility requires constant learning and adaptation from all involved in the software development process.

7.2 Continuous Integration/Continuous Delivery (CI/CD)

Continuous Integration/Continuous Delivery (CI/CD) represents a paradigm shift in software development, moving away from infrequent, large-scale releases to a more iterative and agile approach. It's not just about faster releases; it's about building a robust, reliable system for delivering high-quality software consistently. Think of it as an assembly line for software, automating many of the traditionally manual tasks, and drastically reducing the risk associated with each release.

The core of CI/CD lies in automation. Imagine a scenario where every code commit triggers a series of automated tests, from unit tests to integration tests. This immediate feedback loop highlights issues early, preventing small problems from escalating into major crises later in the development lifecycle. For beginners, starting small is key. Focus on automating a single aspect of your workflow, perhaps just the unit tests, and gradually expand the automation pipeline as you gain confidence and experience.

Implementing CI/CD requires a well-defined process. It begins with developers frequently committing code to a shared repository, often using Git. Then, the CI component takes over: the code is automatically built, and a comprehensive suite of automated tests are run. If the build fails or tests reveal bugs, the team is immediately notified, allowing for rapid identification and resolution of issues. This early detection is crucial; addressing problems early in the development cycle is exponentially cheaper and faster than fixing them later. The successful completion of this stage signals readiness to move to the next step.

The CD (Continuous Delivery) aspect builds upon the foundation established by CI. Assuming the automated tests pass, the code is packaged and prepared for deployment. This could involve creating a deployable artifact, such as a Docker container or a JAR file. This process is often automated as well, streamlining the deployment process and minimizing manual intervention. While complete automation of deployment to production might be a later goal, initial steps could focus on automated deployment to a staging environment for testing purposes, creating a safe space to identify unexpected behavior before reaching end users.

There are numerous tools available to facilitate the CI/CD process. Jenkins, GitLab CI, and CircleCI are just a few examples. The choice of tool depends largely on your project's specific needs, team preferences, and existing infrastructure. Beginners often find it easier to start with cloud-based solutions, as they handle much of the underlying infrastructure management. These platforms offer user-friendly interfaces and comprehensive documentation, making them ideal starting points for learning and experimenting.

The benefits of adopting CI/CD are considerable. Reduced risk is a major advantage; frequent, smaller releases significantly decrease the impact of potential bugs. The automated testing suite catches problems early, minimizing the disruption caused by production failures. Improved team collaboration is another benefit. The CI/CD pipeline fosters a culture of collaboration and shared responsibility, with developers being immediately informed of any issues that their code introduces. This encourages more frequent, smaller changes and prevents any single developer from accumulating too many changes at once.

Yet, CI/CD implementation is not without its challenges. Setting up and maintaining the CI/CD pipeline requires an investment of time and resources. The process itself demands a disciplined approach from the development team. Careful planning and consistent adherence to established processes are crucial for success. The initial learning curve can be steep, especially for teams new to these practices, but the long-term benefits significantly outweigh the initial investment. The advantages in terms of efficiency, reduced risk, and improved code quality will become apparent over time.

Careful monitoring is crucial. While automation reduces manual work, the entire CI/CD pipeline still requires vigilance. Track build times, test success rates, and deployment frequency. Identify bottlenecks and areas for improvement continuously. By carefully analyzing these metrics, you can refine the pipeline, increasing efficiency and reliability. This data-driven approach is essential for optimizing the process and ensuring its long-term effectiveness. Addressing issues proactively prevents minor problems from escalating into major disruptions.

Finally, consider security implications. Your CI/CD pipeline should be treated with the same level of security as your production environment. Protect your source code and credentials diligently. Use strong authentication and authorization mechanisms, and regularly audit your system to identify and address any vulnerabilities. This proactive approach protects against potential breaches and safeguards your sensitive data. Security is an integral part of every software project; CI/CD is no different. Incorporate security best practices into every stage of the process.

7.3 Ethical Considerations in Testing

Software testing, while crucial for quality assurance, carries significant ethical responsibilities. We aren't just checking code; we're safeguarding user experience, data privacy, and even potentially lives, depending on the application. Consider a medical device: a poorly tested system could have devastating consequences. The weight of this responsibility demands a commitment to ethical best practices beyond merely fulfilling technical requirements.

Ignoring potential biases in test design is a serious ethical lapse. For instance, relying solely on data from a homogenous user group can lead to software that fails to meet the needs of diverse populations. This isn't just about inclusivity; it's about ensuring the software functions correctly for everyone, regardless of background or demographic. Failing to consider these factors can result in unfair or discriminatory outcomes, perpetuating existing inequalities. A rigorous testing strategy should actively seek to avoid these biases, aiming for representative data and diverse testing teams.

Transparency in the testing process itself is paramount. Hiding flaws or downplaying negative results is unethical and could expose users to risks. Openly communicating testing results, including limitations and unresolved issues, builds trust and allows for collaborative improvement. This honesty extends to all stakeholders, including clients, developers, and even potential users through clear and accessible reports. Transparency encourages accountability and allows for timely remediation of critical issues.

The ethical considerations extend beyond the software's functionality. Data privacy and security during testing are paramount. Protecting user data during testing, through rigorous anonymization and secure data handling practices, is not just legally mandated but ethically imperative. Breaches of privacy during testing erode user trust and potentially expose sensitive information to malicious actors. Strong data protection protocols must be integrated into every phase of the testing lifecycle.

Furthermore, the pressure to meet deadlines can sometimes compromise ethical conduct. Cutting corners on testing to meet aggressive project timelines is extremely dangerous. Rushing the testing process risks releasing software riddled with vulnerabilities, jeopardizing user safety and data security. Ethical testers prioritize thoroughness over speed, advocating for sufficient testing time and resources to ensure robust software. Prioritizing user safety over expediency should always be the guiding principle.

Finally, responsible software testing requires ongoing learning and adaptation. The landscape of software development, vulnerabilities, and ethical considerations is constantly evolving. Continuous professional development is essential to remain informed about best practices, emerging threats, and evolving ethical standards. Staying current allows testers to adapt their techniques and ethical frameworks to address new challenges. Attending conferences, workshops, reading industry publications, and staying abreast of emerging technologies are vital for maintaining high ethical standards in the field.

www.ingramcontent.com/pod-product-compliance
Lightning Source LLC
LaVergne TN
LVHW020235210926
848481LV00016B/2333
9798349274305